DEC 4 2009

Let's Subtract
Coins

Kelly Doudna

Consulting Editor Monica Marx, M.A./Reading Specialist

Published by SandCastle™, an imprint of ABDO Publishing Company, 4940 Viking Drive, Edina, Minnesota 55435.

Credits
Edited by: Pam Price
Curriculum Coordinator: Nancy Tuminelly
Cover and Interior Design and Production: Mighty Media
Photo Credits: Comstock, PhotoDisc

Library of Congress Cataloging-in-Publication Data

Doudna, Kelly, 1963-
    Let's subtract coins / Kelly Doudna.
       p. cm. -- (Dollars & cents)
    Includes index.
    Summary: Shows how to use subtraction to find out how many coins one has left after paying for various items and looks at different coins, from a penny to a half-dollar.
    ISBN 1-57765-897-3
    1. Money--Juvenile literature. 2. Subtraction--Juvenile literature. [1. Money. 2. Subtraction.] I. Title. II. Series.

HG221.5 .D656 2002
640'.42--dc21

3 1088 1004 7829 4

2002071184

SandCastle™ books are created by a professional team of educators, reading specialists, and content developers around five essential components that include phonemic awareness, phonics, vocabulary, text comprehension, and fluency. All books are written, reviewed, and leveled for guided reading, early intervention reading, and Accelerated Reader® programs and designed for use in shared, guided, and independent reading and writing activities to support a balanced approach to literacy instruction.

Let Us Know

After reading the book, SandCastle would like you to tell us your stories about reading. What is your favorite page? Was there something hard that you needed help with? Share the ups and downs of learning to read. We want to hear from you! To get posted on the ABDO Publishing Company Web site, send us email at:

sandcastle@abdopub.com

SandCastle Level: Transitional

Coins are money.

one penny = 1¢

 one nickel = 5¢

 one dime = 10¢

 one quarter = 25¢

 one half-dollar = 50¢

We use coins to pay for things.

Let's see what we can buy.

Lisa has 8 pennies.

The marble costs 4¢.
4¢ = 4 pennies

How many pennies
will she have left?

Let's subtract.
8 - 4 = 4

The candy costs 8¢.
8¢ = 8 pennies

Sam has 4 pennies.

How many more pennies
does he need?

Let's subtract.
8 - 4 = 4

Brenda has 7 nickels.

The frozen treat costs 20¢.
20¢ = 4 nickels

How many nickels
will she have left?

Let's subtract.
7 - 4 = 3

11

The donut costs 70¢.
70¢ = 7 dimes

Jay has 4 dimes.

How many more dimes does he need?

Let's subtract.
7 - 4 = 3

$1.00

Jen has 6 quarters.

The toy spider costs $1.00.
$1.00 = 4 quarters

How many quarters
will she have left?

Let's subtract.
6 - 4 = 2

The toy airplane costs $3.00.
$3.00 = 6 half-dollars

Lin has 4 half-dollars.

How many more half-dollars
does she need?

Let's subtract.
6 - 4 = 2

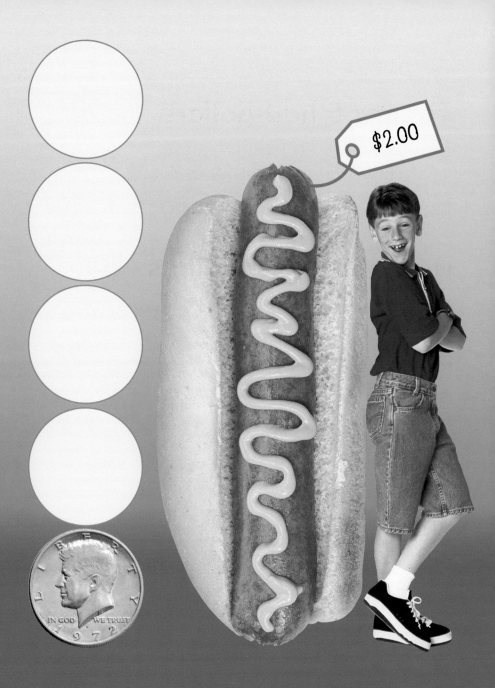

Pat has 5 half-dollars.

The hot dog costs $2.00.
$2.00 = 4 half-dollars

How many half-dollars
will he have left?

Let's subtract.
5 - 4 = 1

What are these coins called?

How much are they worth?

one penny = 1¢
one nickel = 5¢
one dime = 10¢
one quarter = 25¢
one half-dollar = 50¢

Index

Glossary

airplane   a machine that flies and carries people and cargo

donut   a fried, round cake often eaten for breakfast

marble   a small, glass ball used for playing various games

spider   a small creature with eight legs that spins silk for making webs, nests, and cocoons

About SandCastle™

A professional team of educators, reading specialists, and content developers created the SandCastle™ series to support young readers as they develop reading skills and strategies and increase their general knowledge. The SandCastle™ series has four levels that correspond to early literacy development in young children. The levels are provided to help teachers and parents select the appropriate books for young readers.

Emerging Readers
(no flags)

Beginning Readers
(1 flag)

Transitional Readers
(2 flags)

Fluent Readers
(3 flags)

These levels are meant only as a guide. All levels are subject to change.

To see a complete list of SandCastle™ books and other nonfiction titles from ABDO Publishing Company, visit **www.abdopub.com** or contact us at:

4940 Viking Drive, Edina, Minnesota 55435 • 1-800-800-1312 • fax: 1-952-831-1632